John Rollo

Observations on the Means of Preserving and Restoring Health in

the West-Indies

John Rollo

Observations on the Means of Preserving and Restoring Health in the West-Indies

ISBN/EAN: 9783337318680

Printed in Europe, USA, Canada, Australia, Japan

Cover: Foto ©berggeist007 / pixelio.de

More available books at **www.hansebooks.com**

OBSERVATIONS

ON THE

MEANS

OF

PRESERVING and RESTORING

HEALTH

IN THE

WEST-INDIES.

Ye guardian Gods, on whom the fates depend
Of tottering Albion! ————— ———— ————
————— ————— ————— ——— Ye Powers
That o'er th' incircling elements preside!
May nothing worse than what this age has seen
Arrive! Enough abroad ————— ————— ———
Has Albion bled. ————— ———— ————— ———
——— In the West, beyond th' Atlantic foam,
Her bravest sons, keen for the fight, have dy'd
The death of cowards and of common men;
Sunk void of wounds, and fall'n without renown,
      ARMSTRONG's Art of Preserving Health.

LONDON:

Printed for C. DILLY, in the Poultry.

MDCCLXXXIII.

TO THE

OFFICERS of the ARMY

IN THE

WEST-INDIES.

GENTLEMEN,

I Attempted, in a very fhort Addrefs to the Officers who arrived in the Weft-Indies about the beginning of January 1781, to point out the moft obvious caufes which produce the difeafes in thofe countries ;

and

and to fhew that, by a regularity of conduct, it was in their
power to avoid many of. them,
or to render them lefs active.
The Obfervations I then made
are now to be repeated; at the
fame time they are to be en-
larged and confirmed. The
frequent occafion I have had
of regretting the want of attention and care in the periods
of recovery, induces me to
point out the moft eligible fteps
to be taken to accomplifh the
reftoration of health.

THESE Obfervations, Gen-
tlemen, are addreffed to you
in

(v)

in two capacities; as indivi-
duals, and as men having the
direction of others. As indi-
viduals, your feelings and dif-
cernment will furely lead you
to give a ready compliance
with thofe things recommended to you. As men having
the direction of others, your
good-fenfe and humanity muft
force you to obtain a compliance in them. The foldier
under your command is an
object worthy of the moft ferious attention, and you are
bound by every tie to give it.
On your military behaviour

A 3   depends,

depends, in a great meaſure, particularly in the Weſt-Indies, the preſervation and continuance of his health.

In the courſe of theſe Obſervations, many things will be applied to the private ſoldier, which in ſome inſtances will be applicable to you : when theſe happen, I truſt your ſuperior knowledge will render any direct repetition unneceſſary ; it is therefore avoided.

I feel it neceſſary, Gentlemen, to ſolicit the protection

3   of

of that candour and urbanity which so eminently distinguish your character. I feel likewise an earnest desire to persuade you, that the only motive I have in the publication of these Observations, is the firm belief I entertain of the great influence your own conduct has in the business of preserving and restoring health, in yourselves, and in the men under your command.

AND here I beg leave to acknowledge the obligations I lie under to many Gentlemen in the several departments of
the

the Army for the affiftance they have given me. To Mr. STEWART, Director and Purveyor of his Majefty's Hofpitals in the Caribbee Iflands, I am particularly indebted: but his profeffional abilities and extenfive liberality place him beyond any effort of mine to make his virtues more confpicuous. I alfo confefs my debts to other men, who, though remote from my perfonal acquaintance, are well known by their literary works; a confeffion which will be often repeated, as I fhall carefully

fully mark, whenever I have recourfe to their fentiments.

I now take the liberty of fubfcribing myfelf,

With the greateft Refpect,

Gentlemen,

Your moft obedient Servant,

JOHN ROLLO.

Barbadoes,
1782.

C O N T E N T S.

CONTENTS

OBSER.

OBSERVATIONS

ON THE

MEANS of PRESERVING HEALTH

IN THE

W E S T - I N D I E S.

EXPERIENCE has fufficiently fhewn, that the difeafes which appear in the Weft-Indies, whether confidered as peculiar to them, or as proceeding from caufes prevalent in any country, are in general of a very dangerous nature, and are always of an uncertain and precarious termination.

If

If we take a view of the difeafes
as they arife in the Army, we fhall
find that they chiefly originate
from caufes often within our power
to prevent, or to render lefs active.
In the Weft-Indies an army is fubject to difeafes of a different clafs,
if we determine this from their feverity and fatality, from thofe
with which natives, or even Europeans who are fixed in any conftant fituation, are affected : and
this muft proceed from caufes peculiar to the Army. A foldier is
liable to be removed from place to
place, to be expofed to the incle-
mencies of the weather, and to be
employed in every fpecies of hard
labour; befides, he has no variety
of diet, no choice of fituation,
and

and he muſt comply with what‑
ever is directed. Even this is not
all : a ſoldier, in an individual capa‑
city, has frequently bad inclinations,
which cannot be intirely reſtrain‑
ed ; he has feelings and views pe‑
culiar to himſelf, which are diffi‑
cult to be regulated. Theſe traits
of a ſoldier's character in the two
views in which I have preſented
him, as complying with his military
duty, and as acting in a private
capacity, account for the diffe‑
rence of his diſeaſes, reſpecting
ſeverity and danger, from thoſe of
other men. Officers can by ſtrict
diſcipline prevent irregularity, and
a too frequent indulgence of vi‑
cious inclination ; they can alſo,
by an attention to cleanlineſs, to

regular

regular and well conducted meffing, and to fobriety, abate the ra-
pid and mortal tendency of thofe
difeafes to which foldiers are more
immediately liable.

FROM a fuperiority of rank and
circumftances, officers can regulate
their own diet; they can often make
a bad fituation comfortable; and
they can avoid many expofures
which it is impoffible for a foldier
to fhun or to guard againft. However, I am too fenfible that there
are many officers who are not always able to conduct themfelves as
their knowledge and prudence
would direct. Thefe gentlemen,
particularly in the Weft-Indies,
require an indulgent attention from
Govern-

Government; and they undoubtedly merit a double exertion from
thofe more immediately connected
with them. A commanding officer
has it always in his power to act
the father and the friend to the
virtuous fubaltern. By watching
and directing his behaviour in
health, it is probable he may preferve it; by a kind attention to
him in the hour of difeafe, he gains
the efteem of all around; and he
feels the inexpreffible fatisfaction
which conftantly attends the exercife of humanity.

Of

Of the CLIMATE.

A Country differs in climate from the fun's influence, from rain, and from peculiarities of foil and fituation. The Weft-India climate is different from that of Great Britain and North America, from the rays of the fun being more vertical and conftant; from having in general more rain; and from marfhes, woods, or uncultivated ground. From fome of thefe circumftances, Iflands in the Weft-Indies, though at a trifling diftance, vary confiderably in climate, and in refpect of health. Barbadoes and Antigua may be faid to have a different climate from

that

that of St. Lucia and Tobago.
The two former, comparatively
fpeaking, are as healthy as any
fpots in Europe; and the two latter
are quite the reverfe: the difference arifes from rain, marfhes,
woods, and uncultivated ground.
Heat therefore, even although produced by the vertical rays of the
fun, is not the principal caufe * of

* It may be doubted whether the fun's
heat is a caufe of any difeafe except a temporary head-ach, or what is called the "coup
de foleil." Dr. Monro obferves, in his
Treatife on the Difeafes of Soldiers,
vol. I. page 4, That mere heat of itfelf
is not fuch an enemy to health as is generally imagined. This the troops experienced
at Coxheath in the fummer of 1778, &c.
Dr. Naefmith fays, he obferved the fame
thing in voyages to the Eaft-Indies, which
afford the faireft trials of this kind. — Dr.
Lind's Effay on Preferving the Health of
Seamen, 2d edition, note to page 5th.

the

the unhealthineſs attributed to the
Weſt Indies. The circumſtances I
have mentioned as producing the
ſickly alterations, we have in our
power to remove ; at any rate, to
alleviate or reſiſt. The quantity
of rain can be leſſened by clearing
and cultivating the ground; marſhes
may be drained ; and if this is not
practicable, we can ſelect a ſituation on which they have no effect.
Neceſſity may expoſe us to rain,
and to the vapour of marſhes ;
but even then we can by a guarded
conduct partly reſiſt their injurious impreſſions.

In order to be more explicit
reſpecting the climate of the Weſt-
Indies, I ſhall conſider ſeparately
the

the effects of the fun, night air,
rain, and fituation, with the beft
and moft probable means of lef-
fening the prejudicial confequences
of each.

Of the S U N.

THE human body is poffeffed
of a confiderable power to
refift the effects of heat or cold.
Philofophical experiments have
demonftrated that we are capable
of enduring a degree of heat beyond what our feelings could pof-
fibly fuggeft. The application,
however, of artificial heat differs
confiderably from that of the fun,

from

from its being more equally applied. This is evident in the
" coup de foleil," where the fun's
rays are fuppofed to act in a direct
and partial manner. The effects
of artificial and natural heat may
be faid to refemble each other in
one refpect—that if we divert the
vertical rays of the fun, and have
only an equal heat derived from
the warmth of the furrounding atmofphere, nearly the fame feelings
will be produced ; as languor, or
an univerfal wearinefs, an increafe
of perfpiration, and perhaps a
flight head-ach.

It is feldom we cannot divert
the vertical rays, and bring the
heat of the fun to almoft the fame
mode

mode of application and effect as the heat raifed by common fires; therefore, we may take advantage of the power which our conftitution poffeffes of refifting heat, and of preventing any bad effects from the fun's particular influence by a conftant attention to the common means employed. An umbrella is one of the firft things which prefents itfelf; and its ufe is attended with little trouble, though often neglected. The moft proper umbrellas are thofe made of green filk, and of a large fize. A confiderable addition to their ufefulnefs would be foon felt by a double covering of filk, or, what I prefer, a piece of thin dimitty, extending about twelve inches
around

around the top on their infide. A
handkerchief * folded and put under the hat is not only a good
fubftitute for an umbrella, but
with many it anfwers better. I
know gentlemen whom the ufe of
the umbrella does not fecure from
the head-ach when walking or
riding in the fun, yet are defended
from attacks of this complaint in
the fame degree of expofure merely by the ufe of a handkerchief.
The umbrella and handkerchief
may be ufed at the fame time. A
very thin filver plate extending
over the infide of the hat, and covered with dimitty or any cotton
body, is likewife found ufeful ; and

* Black handkerchiefs are improper,
and all filk ones are exceptionable.

it

it may be made to be transferred from one hat to another. Black hats are very improper in the Weſt-Indies, although they are chiefly uſed. Soldiers ought to be allowed to wear white hats, which are procured with eaſe, and not at a dearer rate, I believe, than black. Thoſe who are deſtined for the Weſt-Indies may have their common regimental hats iſſued to them without the black dye; and when they are in the Weſt-Indies, they may be ſupplied from home with the ſame kind. Dr. Lind *, a gentleman to whom the military world is much indebted, obſerves, " that the black hat, which con-

* Eſſay containing Advice to Europeans in Warm Countries, page 250.

" ſtitutes

" ftitutes part of the regimental
" drefs of an Englifh foldier, is
" altogether improper in hot cli-
" mates ; as in thofe countries fol-
" diers are apt, in the heat of the
" day, to be fuddenly feized with
" a fpecies of apoplexy, occafion-
" ed by the fcorching beams of
" the fun, darted on the head, and
" abforbed by the blacknefs of
" the hat ; to prevent which a -
" white covering for that feems
" requifite." If white hats cannot be admitted, foldiers fhould be
directed to have the crown of their
hats externally covered with thick
white paper, and faftened by the
common hat-bands. If walking
and riding in the fun, or any kind
of expofure to him, cannot be dif-
penfed

penfed with, the greateft attention
fhould be paid to the hints I have
given. In riding or walking, the
lefs motion excited the better; for
in proportion to that, the fun's influence will affect.

NATURE, as if confcious of the
effects of heat, has in the Weft-
Indies generoufly provided the re-
frefhing breeze and acefcent fruit.
Art has likewife contributed to
the fame purpofe by the well-adapt-
ed houfe. Thomfon beautifully
expreffes the fhelter Nature gives :

Bear me, Pomona! to thy citron groves;
To where the lemon and the piercing lime,
With the deep orange glowing thro' the
    green,
Their lighter glories blend. Lay me, reclin'd,
Beneath the fpreading tamarind, that fhakes,
Fann'd by the breeze, its fever-cooling
    fruit.

Deep

Deep in the night the maffy locuft fheds,
Quench my hot limbs ; or lead me thro' the
      maze,
Embowering endlefs, of the Indian fig ;
Or thrown at gayer eafe on fome fair brow,
Let me behold, by breezy murmurs cool'd,
Broad o'er my head the verdant cedar wave,
And high palmetos lift their graceful fhade ;
Or, ftretch'd amid thefe orchards of the fun,
Give me to drain the cocoa's milky bowl,
And from the palm to draw its frefhening
      wine,
More bounteous far than all the frantic
      juice
Which Bacchus pours. Nor on its flender
      twigs,
Low bending, be the full pomegranate
      fcorn'd ;
Nor, creeping thro' the woods, the gelid race
Of berries. Oft in humble ftation dwells
Unboaftful Worth, above faftidious Pomp :
Witnefs, thou beft anâna ! thou, the pride
Of vegetable life, beyond whate'er
The poets imag'd in the Golden Age :
Quick let me ftrip thee of thy tufty coat,
Spread thy ambrofial ftores, and feaft with
      Jove !

3
WHEN

WHEN a head-ach, thirſt, or any
uneaſy feeling, ariſes from expo-
ſure to the ſun, it may be generally removed by reſt in the ſhade;
by abſtaining from vinous and ſpi-
rituous liquors; and by the free
uſe of lemonade, cream of tartar
and water, the juice of oranges,
or cold infuſions of tamarinds.
If, however, any of theſe complaints continue more than twelve
hours, gentle evacuation, if not
effected by the preceding drinks,
will be neceſſary; and that may be
procured by a ſmall quantity of
Glauber ſalt.

C   Of

BY night air I comprehend that diverfity of air which occupies the fpace from the fun's quitting the horizon to his return in the morning.

THE night air in every country is deemed prejudicial to health, and by the prudent always guarded againft. In elevated fituations; in abodes not infefted by marfhy exhalations; and where the atmofphere is generally temperate and ferene, fanned occafionally by cooling breezes, the night air is lefs hurtful and dangerous. But

3

in

in countries like the West-Indies, where the heat of the fun is intenfe, where there are frequent falls of rain, and where unhealthy fituations appear, the nocturnal air is baneful, and ought to be fhunned. Dr. Monro, who was a long time at the head of the hofpital department in the army, obferves *, " that nothing has been " found to be more productive of " difeafes in warm climates than " expofure to the damps, efpeci- " ally lying on the ground after the " dews have fallen." A foldier's duty often expofes him unavoidably to all the extremes of the

* Difeafes of the Army, 2d edit. page 45.

C 2   night;

night ; however, by an attention
to a few precautions, any injurious
effect may be in a great meafure
baffled. In all poffible cafes, foldiers on centinel or other fimilar
duty fhould be fheltered by fome
proper covering—a houfe, compact
hut, tent, or the boxes commonly
ufed. Every foldier on expofed
duty ought to be provided with a
watch-coat, which may be very
eafily carried without obftructing
his fervice ; likewife, woollen
ftockings, whole gaters, and thick
fhoes, are infinitely more proper
than the trowfers now injudicioufly
in ufe : and I here prefer the
thick waiftcoat and breeches to
thofe made with nankeen and
linen,

linen, which are at prefent fubfti-
tuted. In the day-time a light
drefs is comfortable and beneficial;
but it is quite the reverfe in nocturnal duty. Soldiers fhould be
allowed to take with them a fmall
quantity of fpirit, and encouraged
to fmoke or chew tobacco in every
fituation of night duties. Soldiers
ought to be ftrictly ordered not to
reft or lie down on the damp
ground. Men upon outpofts, when
no difadvantage to the fervice can
attend, would feel not only pleafing fenfations, but likewife falutary effects, from collecting wood
and burning it, which warms, and
corrects the furrounding atmofphere.

C 3 In

In the morning and evening, especially in the winter months, we are senfible of a degree of cold exciting chillinefs : this, however, is remarkably different from thofe fenfations induced by a cold air in a northern climate. In the Weft-Indies, it produces languid and difagreeable emotions ; in the other, cheerfulnefs and activity. Thefe feelings point out the warm cloathing ; but officers in general, not aware of the pernicious effects of an air of that kind, mount guard, or do any other duty which expofes them to the inclemency of the night, with the fame cloathing they wear when the fun is in his meridian. Inftead of the whole gater recommended

commended for foldiers on night
duty, officers, as they have it in
their power, fhould conftantly
ufe boots.

THE preceding obfervations
are applicable to rainy periods,
as I fhall immediately explain.

Of RAIN.

THE rainy feafons have always proved the moft un-
healthy, not only in the Weft-Indies, but in every part of the
world. Experience has told us,
that even thefe periods may be
rendered

rendered lefs unfalutary, by an attention to our mode of living,
cloathing, lodging, and fituation.
Our living fhould be free and generous, without intemperance and
irregularity. Our cloathing ought
to be of that kind which is beft
calculated to defend us from the
rain, and beget a natural warmth
which may prevent any difagreeable effect. Our houfe muft not
admit the rain, but be compact,
warm, and dry ; and its fituation
muft be out of reach of the noxious vapour of marfhes, and of the
air which paffes through impene-
trable woods. Part of what I
have faid, Dr. Armftrong elegantly
comprehends in the following
lines :

If

If the raw and oozy heaven offend :
Correct the foil, and dry the fources up
Of wat'ry exhalation.——————
——At home with cheerful fires difpel
The humid air ; and let your table fmoke
With folid roaft, or bak'd, or what the herds
Of tamer breed fupply.——————
Generous your wine, the boaft of rip'ning
      years,
But frugal be your cups ; the languid
      frame,
Vapid and funk from yefterday's debauch,
Shrinks from the cold embrace of wat'ry
      heavens.

THE directions I have given on
duty, under the article of Night
Air, are here equally introduced
and recommended. Soldiers when
they are relieved, after being expofed to rain, fhould be ordered
inftantly to their barracks or apart-
ments, whatever they are, to fhift
themfelves, and previous to going
into

into bed to kindle fires either within or at the doors of their habitations; and during this they may take a glafs of fpirit, and indulge in fmoking or chewing tobacco. Dr. Monro fays, " that in wet " weather centinels, or men upon " outpofts, fhould have a fmall " glafs of pure fpirit given them in " prefence of the officer or ferjeant " of the guard : fires in the rear " of the camp for men coming off " duty to warm and dry themfelves " at, were found to be of great fer- " vice." Mindererus recommends *, " in wet, unwholefome " feafons, to make fires of wood

* Medicina Militaris, chap. iv. Englifh tranflation, page 23.

" before

" before the tents, and to burn
" wetted gun-powder, to purify
" the air :" and he obferves, " that
" volleys of fhot made mornings
" and evenings in a camp, conduce
" very much to the difpelling of
" mift and qualifying raw air."

In the rainy feafons, and in the night, every duty fhould be dif-penfed with that is not abfolutely neceffary, from the prefence or immediate apprehenfions of an ene-my. By attention to this circum-ftance alone in the Weft-Indies, the lives of many foldiers may be prolonged, to the important fer-vices of their country. Of what confequence is an officer's charac-ter? He is intrufted with the lives
of

of numbers, and is anſwerable for them ; if not oſtenſibly to the public, to the natural feelings of humanity. What a field opens to the benevolent and generous ! A conſtant opportunity is given for the liberal exerciſe of every tender ſuggeſtion. The man who ſaves one valuable life to his country, is a more reſpectable and worthy member of ſociety than he who has deſtroyed any number of its enemies.

Of

Of SITUATION.

ON the choice of situation of the spot on which we fix our residence, though it be but temporary, principally depends the preservation of health in the West-Indies. All countries have their unhealthy places, and these observation has shewn to proceed from marshes, stagnating water, and woods. The most unhealthy country has its healthy situations: even the unfortunate Island of St. Lucia is not without them. But, as we have observed in another place, soldiers cannot always select the spots on which they may fix

their

their abode. When active operations are carrying on, either in defence or attack, it is impoſſible that the healthy or ſickly ſtate of any ſituation can be attended to, every conſideration giving way to ſecurity or ſuccceſs. The moment however in which we are free from danger, the health of the army becomes the principal object; and the firſt thing to be obſerved, is the particular parts where ſoldiers are to be encamped or ſtationed.

It will be unneceſſary to prove that the neighbourhood of marſhes and thick woods is dangerous, it being evident to every one acquainted with the Weſt-Indies : I ſhall therefore content myſelf with ſpecify-

specifying the healthy places, and marking the most probable means of preventing the effects of a bad situation, when necessity admits of no other.

Dr. Lind * says, " experi-
" ence fully confirms this truth,
" that in such elevated and tem-
" perate situations, where the
" soil is dry and gravelly, and clear
" from wood, shrubs, or stagnating
" water, Europeans enjoy good
" health in the hottest climates,
" during all the seasons of the
" year."

* Advice to Europeans in Hot Countries, page 219.

Dr.

DR. MONRO ✝ obferves, "that
" the moft healthy fituations in
" warm climates are thofe on
" the fides of hills or mountains,
" where the foil is dry, and clear
" from woods and ftagnating wa-
" ter, and where there are no mo-
" raffes within three miles."

THESE quotations fufficiently
point out the places to be felected
for the refidence of officers, and
for the encampment of foldiers.
I fhall only add, that the windward parts of an ifland, and thofe
moft expofed to a wind not inter-
cepted by woods, or impregnated

✝ Difeafes of Soldiers, 2d edit. page 45.

with

with any thing from the land, are the moſt healthy ſituations.

AFTER a healthy place is found, and officers and men are fixed, I ſincerely adviſe them not to go even on a viſit, except obliged by the moſt poſitive duty, to an unhealthy ſpot. Dr. Lind mentions an inſtance of ſome officers whoſe uſual reſidence was on Monk's-hill, from ſleeping a night or two in Engliſhharbour, Antigua, being attacked with a yellow fever on their return to that healthy eminence. More inſtances of this kind could be adduced, but the preceding one is ſufficient; and I obſerve that this and all other ſimilar caſes are applicable to Morne Fortune and the town of Carenage, St. Lucia.

D　　　That

That Ifland has proved fatal to
many officers and men ; and on, a
clofe enquiry, we fhall find the mif-
fortune can with certainty be chiefly attributed to a wandering from
a healthy fituation to one perfectly
oppofite. Officers may trace ficknefs and death from the time fpent
in the Carenage, which is a collec-
tion of houfes furrounded by mor-
tality. Officers who have attached
themfelves to their home, while
they lament the imprudence of
others, feel the heart-felt fatif-
faction of a prudent conduct, and
that good ftate of health which
feldom fails to attend it. Here I
remark, that the marfh furrounding one fide of the Carenage,
might be drained by cutting ca-
nals,

nals, &c.; and I think this is a
fcheme which deferves the attention of a commander in chief.

When the neceffity of fervice
deprives us of choice, and com-
pels us to fix upon the moft unhealthy fituation, we are to make
ufe of the means which are beft
calculated to prevent its unfalutary
effects. When we are in the neighbourhood of marfhes, and to leeward
of them, we fhould have that fide
of the houfe or hut which faces
them fhut up as clofe as poffible,
and the door and windows made in
the oppofite fide. If an officer has
a marquee, the front of it fhould
be placed from the marfh, and the
back part towards it; the fame

D 2   thing

thing ought to be obferved in the encampments of the men. In thefe fituations, wood fires between the marfh and the hut or tent, twice or thrice a-day, particularly in the morning and evening, and frequently made during the night, would be attended with advantage : fmoking or chewing tobacco is likewife found ufeful. An infufion of bark, fnakeroot, or any bitter, in fpirit, has been alfo recommended : a little of either by itfelf, or mixed with water, taken in the morning or when expofed in the night, may be ufed. If a bitter infufion cannot be procured for the men, a fmall quantity of common fpirit fhould be given to them, under the infpection of an officer,

officer, or of a trufty non-commiffioned officer. When guards are fixed on unhealthy fituations, each man
fhould take, on mounting, a large
dofe of powdered bark in water
with a little fpirit ; and this fhould
be repeated when he is relieved.
Every man coming from fuch fituations fhould be examined, and if
he has the flighteft head-ach or
ficknefs, an emetic immediately
given, and followed by two or three
dofes of bark, might prevent a
ferious attack of the difeafe. By
attending to thefe things, the furgeon's and officer's trouble will be
amply rewarded. If the guard
kept at the Carenage in St. Lucia
was treated according to the preceding rules, the danger which at

D 3   prefent

prefent attends that fervice would probably be obviated. Thefe re-marks, which I have made on the means of preventing the bad effects of an unhealthy fituation, throw opportunities in the way of officers to fhew the care and regard they have for the men belonging to them. Soldiers, however inattentive, are not infenfible of kind offices ; they will repay their officers by fidelity and conftancy in danger : at any rate, the pleafure of having merited them is impreffed.

I shall add the fentiments of Dr. Lind * refpecting the prevention of difeafe from unhealthy fituations.

* Advice to Europeans in Hot Countries, pages 149, 151.

THE

" THE beſt preſervative againſt
" the miſchievous impreſſion of
" a putrid fog, a ſwampy or
" of a marſhy exhalation, is a
" cloſe, ſheltered, and covered
" place ; ſuch as a houſe in
" which there are no doors or
" windows facing theſe ſwamps;
" If, in ſuch places, a fire be kept
" either in the chambers, or at
" the doors, or other inlets into a
" houſe, (as is practiſed in ſome
" unhealthy countries, during the
" rainy or noiſome foggy ſeaſon)
" theſe fires, together with the
" ſmoke, prove an excellent and
" effectual protection to thoſe
" within againſt the injuries of a
" bad air. Swampy foreſts emit

D 4 " putrid

" putrid vapours, which are apt to
" produce an immediate ficknefs,
" a vomiting, and afterwards a
" low nervous fever. In fuch cir-
" cumftances, a vomit taken im-
" mediately, and a change into a
" pure air, will often prevent a fit
" of ficknefs."

ARMSTRONG fays :

————I praife the man who builds
High on the breezy ridge, whofe lofty fides
Th' ethereal deep with endlefs billows
    chafes.
His purer manfion nor contagious years
Shall reach, nor deadly putrid airs annoy.
But may no fogs, from lake or fenny plain,
Involve my hill ! And wherefoe'er you
    build,
Dry be your houfe; but airy more than
    warm.

————On

————On the marſhy plains
Build not ; nor reſt too long thy wand'ring
feet.
For on a ruſtic throne of dewy turf,
With baneful fogs her aching temples
bound,
Quartana there preſides.

——— Avoid the mournful plain
Where oſiers thrive, and trees that love the
lake ;
Where many lazy muddy rivers flow :
Nor for the wealth that all the Indies roll,
Fix near the marſhy margin of the main.

Of

Of EFFECTS *peculiar to* the WEST-INDIA CLIMATE.

STRANGERS after their arrival in the West Indies are liable to fome complaints which may be faid to be peculiar to the climate : and thefe do not feem to arife diftinctly from any one of the circumftances which I have already confidered ; but they appear to proceed from a conjunction of fome of them, or rather from fomething not well afcertained. Every perfon, however guarded and cautious even in thofe places deemed the moft healthy, is fubject to the complaints alluded to, and feldom or ever efcapes an attack of them. They are, however, more troublefome than dangerous,

rous, yielding always to a mild treat-
ment, and I may fay never proving
fatal, if early and judicioufly attended to.

HEADACH with ficknefs or loath-
ing of food, a bitterifh tafte, coftivenefs, and a high-coloured urine,
are among the firft things which
affect Europeans. Thefe foon go
off by confinement, an abftinence
from the ufual diet and liquor, and
a free ufe of lemonade, cream of
tartar and water, or tamarind be-
verage. If they fhould continue
after this twenty-four hours, gentle
evacuations, &c. will be neceffary;
but here recourfe muft be had to
the furgeon.

ERUPTIONS, as the prickly-head,
and another kind generally taken
for

for mufquetoe bites, are the conftant
attendants upon all new-comers.

THE prickly-head is not altogether confined to ftrangers, it vifits
fome of the natives annually : it is a
mere external eruption, and not
connected in itfelf with any other
affection. It is a vulgar idea that
it is a falutary appearance, and is
thrown out from the blood ; it
affects only the fkin, and is produced, I think, principally by the
action of heat upon it. What has
given rife to its being a more ferious
thing, is the common obfervation,
that when it difappears a head-
ach, &c. attend. This is not always the cafe ; and when it does
happen, the headach and other
complaints generally precede its
disappear-

difappearance, which is only produced by the intervention of an
accidental difeafe, that diminifhes
or contracts every exterior part.
A variety of external applications
are recommended, but I never faw
one of them repay the trouble ac-
companying its ufe. A light cool
drefs and patience are the only
things I advife.

THE other kind of eruption,
which is often taken for mufquetoe
bites, requires more attention;
for it is frequently accompanied
with fever, generally with head-
ach, which in many cafes is attended with an inflammation of
the eyelids. This eruption commonly appears firft on the legs and
thighs, then on the arms, neck, and

rent

face, refembling bumps of different magnitudes, from the fize of
a pea to four times its bignefs:
they have an extenfive bafe, and
protrude above the fkin, terminat-
ing in an apex or point. They
itch exceedingly, and it is next to
impoffible to refrain from fcratching, which is fure to make them
bleed, and increafe the inflammation and pain. They continue for
fome days, then gradually difap-
pear, and are fucceeded by a new
fet; which is in many inftances
again repeated. If a headach and
inflammation of the eyes, or any
mark of fever attend, directions
fhould not be trufted to, but medical affiftance be called in. For the
eruption itfelf, all that is neceffary
confifts in a moderate and light
diet,

diet, a plentiful ufe of lemonade
or tamarind drink, which keeps
the belly gently open, and in re-
fraining from fcratching as much
as poffible.

DR. HILLARY, who treats of the
difeafes of Barbadoes, takes notice
of both thefe eruptions, and gives
an accurate defcription of them.
He obferves, that they often appear
at the fame time, by which every
fymptom is more troublefome.
He condemns the ufe of external
applications, particularly in the
prickly-head, which he fuppofes
may repel it, and produce dangerous effects. It is immaterial,
in common directions, whether
we fay external applications repel
the prickly-head, and bring on
more ferious difeafes ; or that they

are

are more troublefome than ufeful,
therefore may be altogether laid
afide.

I shall here introduce a method, which I am confident, if it
was ftrictly followed, would be attended with the beft effects, in preventing foldiers from being attacked with difeafe on their arrival in
the Weft-Indies. I tried it on a
detachment of the Artillery, confifting of twenty men, who came from
England, after they were landed at
Barbadoes ; and I am certain, advantage was the confequence.
The trial muft, I allow, be carried
further; and it may be extended to
other fituations, which I fhall point
out. The method is not my own;

it

It comes from refpectable authori-
ties, but it is not profecuted; at
leaft, if it is or has been in the Weft
Indies during this war, the practice
muft be partial, and not fo generally
ufed or known as its importance
merits. The whole bufinefs is
comprehended in giving the men a
certain quantity of Peruvian bark
once a day, and repeating it for a
few more; then leaving it off two
or three days; commencing again,
and continuing it for three or four
days longer; after which it is to be
difcontinued. Each man in this
way is fuppofed to take about two
ounces of the bark, which to a
regiment confifting of fix hundred
men will amount to feventy-five
pounds weight. This quantity of

E   bark

·bark given in the manner I direct, will probably save more than three times the weight.

THE Artillery men on whom I began this practice at Barbadoes, were paraded in the morning about eight o'clock; their number was twenty. I mixed in a vessel forty drachms, which is equal to five ounces of powdered bark, with four pints of common water, and half a pint of rum. Of this mixture, I gave out of my own hand to each man nearly one gill, whch disposed of the whole quantity. This was repeated at the same hour for two or three mornings, then discontinued; renewed again in three or four days, and continued until each soldier had taken two ounces of bark, when it was entirely left off.

3   BEFORE.

BEFORE I attempted this practice, the men began to complain daily of bilious affections; but after they had taken the bark, thofe complaints ceafed, and the men continued in tolerable health, although they were unavoidably put upon fatigue in difembarking and arranging ordnance ftores. During the preceding exhibition of the bark, the greateft attention was given to regularity and cleanlinefs, and the men who had no complaint bathed in the fea every morning before the hour of parade. But of thefe I fhall hereafter have occafion to make more particular mention.

IF the practice I have recommended was rigidly attended to in more unhealthy iflands than Barbadoes,

E e   the

the advantages would be more con-
spicuous, and fully repay any expence or trouble. It must appear
at first view, that the medical character is not altogether the acting
one, here; the officer holds a dif-
tinguished part, and without his
affistance very little can be done
or expected. Soldiers do not confider themselves under the furgeon's
direction until they are fick; here
only men in health are concerned;
therefore the officer muft interfere,
and infpect the taking of the medicine. This is a duty of humanity in which all the generous
feelings are interefted; the officer
will therefore exert his influence.
In the Weft-Indies we muft go
hand in hand; for to me it is fcarce
a mat-

a matter of doubt, whether the officer is not more concerned in the prevention of ficknefs among the foldiers, than the furgeon.

It will be unneceffary for me to point out from what fource the bark is to be fupplied, or to mark the particular modes which corps may adopt in giving it, thefe naturally occurring to thofe of whom it is required. I humbly folicit the attention of the commander in chief, the commanding officers of regiments, and the gentlemen at the head of the hofpital department, to what I have taken upon me to recommend ; and I alfo beg a continuance of it, to the obfervations which immediately follow.

E 3

As

As it is certain that some of the islands in the West-Indies are more healthy than others, is it not to be presumed that eminent ad- vantages would accrue to the ser- vice, if soldiers, on their arrival in these countries, were stationed for some time in the most healthy islands, before they were sent to those which are deemed more un- healthy? In the one situation, they have only the effects which I have marked as peculiar to the West-In- dia climate to encounter; in the other, they have these united with causes that never fail to produce dangerous diseases.

IMMEDIATELY after a man's arrival in the West-Indies, if he is placed on an unhealthy spot,

ſpot, he is attacked with a bilious complaint, which in a
healthy ſituation would not be dangerous ; but here it becomes ſo,
becauſe it renders the body accef-
ſible to the effects of the unhealthy
ſpot on which he is fixed. If this
man on his arrival in the Weſt-Indies was placed on a healthy ſituation, he would probably have an
attack of bilious complaint, but
ſuch a one as would ſoon leave him.
After this, 'tis ſix to one if a removal to an unhealthy ſituation
would produce another viſit of the
bilious complaint ; for which rea-
ſon he will not be ſo liable to be
affected by the cauſes of diſeaſe
that ſurround him. However, I
by no means advance, that ſuch a
man will not have a dangerous diſ-

eaſe

eafe in the unhealthy fituation ; I
only infer, that the probability on
the other fide is in his favour.

FROM what I have faid I wifh to
convey this obfervation, that all regiments or detachments of men
from Europe, on their arrival in the
Weft-India Caribbee Iflands, fhould
be ftationed in Barbadoes or Antigua until they are habituated to
the nature of the climate ; then
they may be fent to more unfavour-
able fituations, being previoufly
relieved by a fimilar fet from England, or by men from that place
to which they are deftined. By
this ftep two purpofes are anfwer-
ed : Men have a fairer chance of
refifting the difeafes in the Weft-

Indies; and men who have been
fome time in an unhealthy fituation, and very likely nearly worn-
out by difeafe, are relieved, and
have a profpect of once more en-
joying tolerable good health.

Of LODGING.

I HAVE already obferved the bad
effects of the Weft-India climate under the circumftances of
the fun's influence, night air, rain,
and unhealthy fituations, which
fufficiently point out the kind of
lodging moft fuitable to prevent
them.

them. I fhall therefore only further obferve, that after a choice of
fituation, it ought to be the next
object with the commanding officer, to fee that his inferior officers
and men are comfortably fheltered
in cool and dry lodgings. Here I
might fummon the attention to
prove the prejudicial confequences
which follow a leaky roof, and a
damp floor; but it will be quite
enough to found the fatal name
" St. Lucia !" It is impoffible to direct our eyes towards that unfortunate ifland, without lamenting the
fate of many valuable lives,
which have fallen facrifices to an
ill-concerted œconomy, or fome
other miftaken fyftem. Well may
we

we at this day adopt the language
of Armftrong—

Albion's————
—— braveft fons, keen for the fight, have
    dy'd
The death of cowards and of common men ;
Sunk void of wounds, and fall'n without
    renown.

AFTER comfortable and well-
placed lodgings are provided, the
next thing which occurs is the management and behaviour of foldiers in them. Soldiers are apt
to lounge and loiter in their barracks, which never fhould be permitted, as indolence is not only
prejudicial in itfelf to health, but
more fo by begetting dirtinefs and
filth. Certain regulations fhould
be adopted, to keep, on pain of
                        punifh-

punifhment, the lodgings clean and dry, and free from incumbrances. If poffible, nothing ought to be admitted within the barrack but what is abfolutely neceffary : all fpare cloathing and accoutrements fhould be fome other way difpofed of. A foldier may be allowed to eat his victuals there ; but after doing this, the place ought to be carefully fwept, and the utenfils of the mefs quickly cleaned and put away. Wafhing the face and hands and linen, combing the hair, brufhing cloaths, cleaning belts and firelocks, are to be forbidden, except in circumftances where it is impoffible to avoid doing thefe things within the barrack ; and then double care fhould be beftowed in fweeping,

fweeping, and preventing a reten-
tion of moifture. The bedding,
of whatever kind, muft be aired
every dry day ; and if it confifts
of any thing that can be wafhed,
it ought to be wafhed once or twice
a week. Any difficulty of having
thefe things ftrictly performed, lies
only in appearance ; on trial, it
immediately vanifhes, depending
entirely on orders being given and
obeyed, which every officer can
always accomplifh. I have pur-
pofely omitted faying any thing
about neceffary-houfes, the propriety of having them in the rear
of lodging or encampment, and of
keeping them clean, being obvious
to all.

Of

Of DRESS.

CLEANLINESS is not only pleafing, but comfortable; it is falutary and beneficial. A foldier cannot be too rigidly attended to in point of drefs; for the one who is conftantly neat and clean, is neither indolent or lazy; two things which I have obferved are prejudicial to health. In the Army, every thing ought to be done by rule. At an allotted period, foldiers fhould comb their hair, wafh themfelves, and put on their cloaths; and thefe are to be performed under the infpection of non-commiffioned officers; and if they

require

require notice, this is a duty not beneath a more elevated character. In the day-time foldiers may be indulged in the choice of what kind of waiftcoats, breeches, trowfers, and ftockings, they will wear ; but in night duty, or in cafes of expofure to rain, they fhould be only permitted to wear the regi-mental waiftcoat, breeches, and ftockings. No material inconvenience can arife from the change of drefs which here takes place. Dr. Monro fays, " Soldiers fhould be " obliged to keep themfelves neat " and clean ; to comb their hair, " and change their linen often ; " and if the camp be near the fea " or a large river, they ought to " bathe themfelves early in the " morning,

" morning, as the fervice will
" permit."

BATHING in cold water in the
Weſt-Indies being followed by
cheerfulneſs and activity, becomes
very neceſſary, beſides its being
conducive to cleanlineſs. Officers
in any ſituation can have almoſt
every advantage to be derived from
plunging in the ſea, by having two
or three pailfuls of cold water
thrown over them by a ſervant
early in the morning ; or if conve-
niency will allow, the ſhore bath
may be adopted. Soldiers ought
not to have a diſcretionary power
to bathe, becauſe they are liable
to abuſe it, either by chuſing an
improper time, or being in a ſtate
in

in which it may prove hurtful. If regiments or detachments were paraded at a felected hour, and marched off to bathe under the direction of officers, thofe men whom the furgeon judged to be improper fubjects being previoufly excluded, then the good effects of bathing, without its inconveniences, might be expected. The detachment of Artillery formerly mentioned as having bathed while they were ufing bark, immediately after their arrival at Barbadoes, went into the fea two hours before the morning parade ; which made the hour of bathing to be about fix o'clock, the period I judge moft proper.

THE circumftances which render bathing improper, are in-

F   difpofition

diſpoſition of any kind, except
ariſing from weakneſs, (but here
the ſurgeon interferes, and muſt
determine) headach, and the prefence of eruptions.

SOME doubt ariſes whether the
prickly-heat makes bathing dangerous. I have already ſaid, that I conſider the prickly-heat to be a mere
affection of the ſkin, unconnected
with any other bodily indiſpoſition ;
therefore, I can ſee no injury ariſing
from bathing with it. I have
known many bathe with the prickly-heat, and never ſaw any inconvenience ariſe, except its becoming
more plentiful and troubleſome.
However, bathing with the prickly-heat is ſeriouſly condemned by
many medical gentlemen in the
West-

Weft-Indies ; Drs. Hillary and
Monro forbid it ; therefore, I
would by no means particularly
recommend or advife it. One cri-
terion may guide us—that if we
bathe with the prickly-heat, and
find it followed by any thing difagreeable, we can defift ; but if it
is fucceeded by only an increafe
of the eruption without any complaint, I think we may fafely
continue.

In this place I cannot omit ac-
knowledging, befides what I have
elfewhere declared, and which I
muft always repeat from the conftant occafion given, the attention of
Major WILLIAMSON, command-

ing

ing the Artillery in the Weft-Indies.—Ever ready to direct what
is neceffary, and to comply with
what is pointed out for the advantage of his men, he has become
the father of the foldier, and the
valuable officer of his country.

Whilst I do juftice to this character, it is impoffible to refrain
from paying a like tribute to Col.
Cuyler, who commands the 55th
regiment. This regiment, for
management and difcipline, gives
the model which every other
corps fhould imitate. The conduct of foldiers in barracks, in
point of drefs and regularity,
and the mode of bathing, are
here inculcated to us by an invariable practice ; and the practice

tice has been repaid by a fuperior degree of healthinefs. From this I by no means infer, that all other regiments are inattentive to regularity and difcipline; I am perfuaded of the contrary: however, I do not hefitate to fay, that more attention might be paid to them in every corps in the fervice.

F 3   Of

Of DIET.

THE diet of officers may be conducted by the following line —

Live well, and live regularly.

Living well and living regularly are far from being incompatible. A tasty and nourishing diet, even a generous allowance of wine, may not only be used, but are absolutely necessary to answer the purposes of health *. The moment,

---

* Dr. Wind, in his notes to a translation of Dr. Lind's Essay on the Diseases incidental to Europeans in Hot Countries, observes, that at Middleburgh, the capital
of

ment, however, we go beyond the cheerful glaſs, that inſtant we ex-poſe ourſelves to every cauſe capable of producing diſeaſe.

FROM a comparative view of the different degrees of health and ſick-neſs among thoſe who have lived in conformity to the maxim ſtated, and in a manner diametrically op-

of Weſt Zealand, in the month of Auguſt, after the rains which happen in July, intermitting fevers prevail: he ſays, ſuch as live well, drink wine, and have warm cloathing and good lodgings during the ſickly ſeaſon, do not ſuffer ſo much as the poor people. Dr. Knox told Dr. Monro, that laſt war, in the Guadaloupe expedition, he obſerved that thoſe who had opportunities of drinking Madeira and claret, and uſed thoſe liquors in moderation, were leſs liable to dyſenteries and bilious fevers than others.

F 4　　　　poſite,

pofite, the beneficial confequences of the one and the pernicious tendency of the other are well af-certained. A vigorous and active conftitution has the greateft proba-ble chance of refifting the caufes of difeafe, in any poffible fituation; therefore, whatever gives and fe-cures that conftitution, is beft cal-culated to preferve health.

A REGULAR and temperate mode of life, a comfortable lodging, a cheerful ftate of mind, and a gene-rous diet without the fmalleft degree of excefs, conftitute the effential parts of the fyftem which feems to me beft adapted to fecure health in the Weft-Indies. This will ap-pear more juft, by a review of the effects of a different manner of life.

EXCESS

Excess of any kind, but more particularly of drinking, produces a certain difpofition of the body favourable to the operation of the caufes of difeafe which I have pointed out, and againft which I have endeavoured to put you on your guard, viz. the fun's influence, night air, and unhealthy fituations. Thefe caufes feldom act alone ; they are generally conjoined, and affift one another ; and they require a certain ftate of the conftitution, before they can produce any fevere or fatal effect. This conftitution is a relaxed and weakened deviation from the natural ftate, and is always attended with a lownefs of fpirits, particularly when it is induced by intemperance

perance and irregularity. Thefe
are truths well known to thofe who
indulge in exceffes of that kind;
for they are conftantly the morning vifitors after a debauch. A
perfon with thefe morning feelings, if expofed to rain, or to the
vapour of marfhes, it is exceedingly probable will be attacked
with a fatal fever. To make thefe
obfervations more convincing, I
fhall infert the fentiments of men
always credited.

Dr. Lind fays*, " that excef-
" five drinking, and every fpecies
" of intemperance, difpofe the

* Advice to Europeans, page 8.

" conftitu-

" conftitution, more efpecially in
" hot climates, to the attack of the
" epidemic difeafes of the coun-
" try." In another place * he
obferves, " that in an air noxious
" from marfhes, or in the unheal-
" thy feafon, any debauch or
" drunkennefs will often give a
" fever, which in lefs than forty-
" eight hours will terminate in
" the death of the patient. Ex-
" ceffes either in eating or drinking,
" in hot countries, are extremely
" prejudicial to the conftitution."

Dr. Monro † is of opinion,
" that nothing has been found to

* Advice to Europeans, pages 186 and
187.
† Difeafes of Soldiers, page 45.

3   " be

" be more productive of difeafes
" in warm climates, than indulg-
" ing freely in the ufe of fpirits and
" other ftrong fermented liquors."
Hence he recommends temperance
in drinking, and particularly
condemns the too liberal ufe of
wine. But

We curfe not wine ; the vile excefs we
   blame,
More fruitful than th' accumulated board,
Of pain and mifery.———
       ARMSTRONG.

EVEN although confcious of the
impropriety of intemperance, it
will happen fometimes that the
moft vigilant will be furprifed into
it ; for when we chance to fall into
a friendly and focial circle, ani-
mated by the mirth and good-hu-
       mour

mour which reign around us, we
are apt to forget ufual reftraints,
and pafs imperceptibly beyond the
limited glafs. In cafes of this
kind, when headach and dejection
of mind are the confequences, and
continue longer than ufual, with
the acceffion of other marks of in-
difpofition, it may with great probability be fufpected, that a difeafe is
forming, from a co-operation of
the caufes of difeafe I have
formerly enumerated ; therefore,
the furgeon fhould be immediately
confulted ; and if any delay occurs
from diftance or otherwife, an emetic of ipecacuanha will be proper,
which all gentlemen on feparate
duty ought to have in poffeffion :
for difeafes in this country are
often prevented by an early exhibition

hibition of a medicine of this nature, and a fuitable reftriction in point of diet.

IT is a good rule, to take the morning fucceeding a debauch, two tea-fpoonfuls of powdered bark in water, which may be repeated once or twice in the courfe of the day ; but this is only to be done when there is little headach, or apparent reafon to think neither that or any other uneafy feeling will continue.

I CANNOT omit here condemning, as big with the moft ferious confe- quences, the idea of keeping off a difeafe by launching into an extreme of excefs, particularly in the article of drink. If this con- duct

duct does not anfwer the intended purpofe, the uneafy feelings which induced it increafe, and form dif- eafe; and this difeafe, inftead of being of a milder kind, will un- doubtedly prove of a more dan- gerous nature than it otherwife would have affumed : whereas, if a fuitable reftriction is made, the difeafe may be either prevented from forming, or, when formed, have its ufual feverity abated.

THESE obfervations refpecting intemperance and irregularity are applicable to foldiers, indeed more fo than to officers; for having lefs variety of diet, no choice but a de- voted ration, they are more ex- pofed to the confequences. Be- fides, foldiers probaby have a
more

more indifferent and lefs comfortable lodging, and from duty and
other circumftances are more liable
to fuffer from the inclemency of
weather, or unhealthinefs of fituation. From all thefe confidera-
tions they ought to be particularly
reftrained from indulging in exceffes of any kind.

THE effects of too great a quantity of wine are lefs hurtful than of
rum : the latter, befides the general
bad confequences of drunkennefs,
acts in a moft injurious manner on
the delicate fibres of the ftomach and
bowels. I knew a young man of
the Artillery at St. Lucia, of a gay
and lively difpofition, who joined
in the practice of drinking a little
pure rum in the morning : this he
continued,

continued, increasing the quantity; until he was carried off by fever and loosenefs. On diffection, his stomach was found ulcerated, and otherwife very fingularly difeafed.

Rum I conceive to be a very ufe-ful article in a foldier's allowance, particularly in the Weft-Indies; but he ought never on ordinary occafions to drink it unmixed. If men cannot be trufted, their rum fhould be ferved out, diluted with four or five times its proportion of water; according to the ftrength of the fpirit. With refpect to the quality of rum, the older it is the better; for in its new ftate it con-tains an acrid corroding principle, which in time evaporates and al-together difappears. This is evi-

G   dent

dent on examining the different ages
of fpirit : the new has a pungent
fmell, and a fharp difagrecable
tafte ; the old has a pleafing aro-
matic flavour, and a mild oleagi-
nous tafte. The rum fupplied to
the army, from what caufe I do
not determine, is generally of the
moft indifferent quality. Commanding officers can condemn in-
jured provifions; may not they
equally fet afide rum of a very new
and bad quality ? Government
gives every care to its foldiers, and
grants every indulgence to them;
but by fome fatality or other, its
intentions are too often proftituted.

Soldiers have methods of pro-
curing rum, above what their ordinary allowance (which is perfectly
fufficient)

fufficient) entitles them to. Thefe
practices can be nearly removed
by a regular and well-conducted
meffing : this attended to in one
regiment, would give it a very
great fuperiority over any other,
and would be the beft means of en-
furing fobriety, decency, and health.

SOLDIERS, as it is the military
practice, though not always ftrictly
followed, fhould be divided into meffes, each under the direction of a non-
commiffioned officer, or a felected
private man, and the whole infpect-
ed by a commiffioned officer. Dinner
is the principal meal; therefore, the
chief attention is to be beftowed
on it. All men not on guard ought
to have a ftated hour, at which
time the officer fhould go the

round, and fee that every body is
at dinner, and that the whole of
the victuals are dreffed. and well-
cooked. Thofe men who are on
guard fhould have, if poffible, their
dinner fent to them, from the re-
fpective meffes to which they be-
long. At firft view this is a troublefome duty, but it is a ftanding
order in the army, and may be ex-
ecuted in a few minutes. By this
mode an effectual ftop is put to the
fale of provifions, too often prac-
tifed, and which procures the
baneful fpirit.

THE articles that foldiers generally difpofe of are the fmall
fpecies, as peafe, oatmeal, or
rice, which are very effential
things in a ration ; they are vegetable preparations, and are the
beft

beſt aſſiſtants with the bread to make the ſalt proviſion nouriſhing, and prevent any of its effects on the conſtitution which may otherwiſe follow. Would not an allowance of a ſmall quantity of vinegar to each meſs, particularly in the Weſt-Indies, be uſeful? If a ſoldier is permitted to ſell or exchange any part of his proviſions, it ſhould be with the conſent of the perſon who directs the meſs to which he belongs; and even then it ought to be only for vegetables, fiſh, or any thing freſh.

WHAT a pleaſure it muſt give to an officer, to have preſented to him ſo many occaſions of being ſerviceable to men who are too apt

to

to neglect themfelves ! The fine
feelings and views of a gentleman,
the benevolence and generofity
which hold the firft places in his
character, the extenfive liberality
of fentiment, and the perfect knowledge of human nature, acquired
by the beft education, are all fuppofed to center in a Britifh officer.
He requires them all, and he will
find in the performance of his duty,
conftant opportunities of exerting
them.

Of

Of EMPLOYMENT.

THE employment of officers, except in matters of duty, falls under their own direction; therefore they have it often in their power to regulate their actions, as far as these are concerned in the preservation of their health. The first thing which I point out to them is,

" To go to bed early, and rise early."

By a strict attendance to this rule, several of the causes of disease which I have marked are avoided. Besides, a principal intention of nature is fulfilled, by giving the

G 4   proper

proper relaxation to our powers of action, which would otherwise be too much fatigued, and in time rendered defective in performing their ordinary motions. This is the chief reason late hours are improper, even though not accompanied with intemperance ; for whatever weakens or leſſens the vigour of either body or mind is prejudicial, becauſe it makes it more liable to be affected with other cauſes of diſeaſe. An officer on duty cannot attend to the preceding injunction ; but unleſs that is more ſevere than uſual in the Weſt-Indies, he can comply with it three nights out of four ; and by doing ſo he is better enabled to reſiſt any inclemency of night duty, when it does occur.

ALL

ALL kinds of exercife are peculiarly neceffary in the Weft-Indies;
however, it is fo unfortunate, that
we have it but very feldom in our
power to felect the moft ufeful kind,
or even to have an opportunity
of ufing any. The mornings
and evenings are the fitteft periods,
and they give only a fmall proportion of time. Riding and walking
are the two modes of exercife thofe
countries afford, and they can be
ufed only with propriety in the periods I have marked. Bathing in
cold water, if duly profecuted,
greatly fupplies the place of exercife :

It is the pureft exercife of health,
The kind refrefher of the fummer heats.
THOMSON.

But

But of that I have already taken notice, under the article of Dreſs.

As the greateſt part of a Weſt-India day muſt be ſpent in the houſe, every thing ſhould be exerted to engage the attention of the mind, and maintain a rational gaiety and cheerfulneſs. Here an officer enjoys the fruits of a good education. If there is nothing around to afford him entertainment, he can take an inward ſurvey, and find the moſt ſatisfactory amuſement in the contemplation of himſelf, of his views and intentions. Books are naturally pointed out, but a ſelection is not always to be met with ; for being a heavy part of baggage, they are ſeldom carried.

carried. If a fmall and well-chofen
library was procured in each regiment, by the fubfcription of its
officers, and put under the quarter-mafter's care, with the ftores of
the regiment, every officer would
be fupplied with books, and without any trouble or incumbrance;
by which fatisfaction and pleafure
might be always at hand.

It may be faid, that the frequent changing of officers in a
regiment, makes fuch a fcheme
impracticable. Not at all; for
it is only paying (by the perfon
who keeps the accounts of the
library, and whom I fuppofe
to be either the paymafter or
quarter-mafter) an officer on his
leaving the regiment the money
he fubfcribed, and getting the
fame

same sum from the person who succeeds to his place. Besides the original subscription, there should be a small annual sum to repair the library and procure new books.

In such confined scenes as the West-Indies exhibit, how enviable are the feelings of an agreeable acquaintance, and of friendship!

———— social friends,
Attun'd to happy unison of soul;
———— Whose minds are richly fraught
With philosophic stores, superior light;
And in whose breast, enthusiastic, burns
Virtue.

Among these we are to expect

The full free converse of the friendly
    heart,
Improving and improv'd'.

                    THOMSON.
                    SOLDIERS

Soldiers never fhould be expofed, except on unavoidable duty, to
the fun, or to fatigue during his
fcorching heat. The hours of parade, of manual exercife, and of
relieving and mounting guard,
ought always to be thofe in which
the fun has the leaft influence.
About funfet I take to be the beft
time for relieving guards; for this
reafon; the men mount refrefhed,
after the repofe of a day, and will
from that be watchful and alert in
the night : whereas they who
enter upon the duty in the
morning, are fatigued upon the ap-
proach of night; therefore are apt
to flumber and be carelefs; by
which the fervice is not only en-
dangered, but the men are more
liable

liable to fuffer from night air,
or unhealthy fituations. Befides,
the men relieved in the evening,
have the advantages of a natural
night's reft to fit them for the
duties of another day, whilft the
other men are under the neceffity
of fleeping in the day, which never
affords equal refrefhment. I mean
here thofe guards which are only
relieved once in twenty-four hours.

IF from the circumftances of
fervice, foldiers are required to
carry their provifions to any diftance ; or if they are under the neceffity of repairing roads, of build-
ing huts, working at batteries,
or of undergoing any fa-
tigue whatever ; the cool hours
of the day fhould be chofen.

From

From what I have feen, and col-
lected from the obfervation of
others, I am confident that foldiers might be excufed from any
of thefe fevere duties which
I have fpecified. Where the
country does not give negroes
or labourers for fuch purpofes,
it is the intereft of Government
to fupply them.

HERE again I will mention St.
Lucia. I am convinced that the fa-
tigues impofed on the foldiers in
that ifland, have proved more de-
ftructive to them than its natural
evils; for thefe alone could not have
produced fuch general fatality, if
they had not been affifted by the effects of the horrid fervices in which
the men have been employed.

DR.

Dr. Lind *; after giving fome
inftances of the fatality attending
the employment of Europeans in
hot countries in laborious work;
particularly that of cutting down
wood; &c. fubjoins the following
very ftriking obfervation, which
I here beg leave to adopt : " It
" does not feem confiftent with
" Britifh humanity; to affign fuch
" employments to a regiment of
" gallant foldiers; or to a company
" of brave feamen."

When the commander in chief,
and all other commanding officers,
give the proper attention to the

* Advice to Europeans, page 145.

reprefen-

reprefentations of the director of
the hofpitals, phyfician and fur-
geons of the army, although they
may be only founded upon probability : then, and not till then,
every advantage tending to the
prefervation of the health of fol
diers in the Weft-Indies may be
expected.   -

H

OBSERVATIONS

ON THE

MEANS of RESTORING HEALTH

IN THE

WEST-INDIES.

FROM difeafes in the Weft-Indies the recovery is often very tedious and uncertain : this, however, fometimes arifes from neglect and inattention. When a patient is pronounced free from danger, he is too apt to fhake off thofe little reftraints which are abfolutely neceffary to fecure a re-

covery

covery and prevent a relapfe. I have frequently had occafion to lament the injudicious and carelefs conduct of many, even although repeatedly warned of the confequences. I have feen men on the recovery—I have feen them relapfe, and fall facrifices to imprudence and folly.

AFTER the termination of a difeafe, there is a great deal remaining to effectuate a perfect reftoration of health. Every difeafe leaves a debilitated, weakened, or relaxed ftate of the conftitution; and if this is not removed by the natural and artificial efforts for that purpofe, a dropfical, pectoral, or fome other complaint, if not a

relapfe

relapfe into the former difeafe,
will certainly follow. The artificial efforts in reftoring a weakened
conftitution, receive but a trifling
aid from the province of medicine ; they confift chiefly in perfonal attention to good nurfing, activity, and cheerfulnefs, but, above
all, to a change of air.

. It will be unneceffary to ad-
duce inftances in fupport of thefe
obfervations, as perfons recovering from difeafe are perfuaded of
the truth of them, but want for-
titude and refolution to comply.
However, I flatter myfelf, by
pointing out the fteps to be taken
in the ftages of recovery for the
re-eftablifhment of health, mark-

H 3   ing

ing occasionally the difadvantages of a different fyftem, that every one will allow the attention and compliance he may feel himfelf interefted to give. What I have to recommend will be comprehended under the following articles—Change of Air, Diet, Drefs, and Employment; to which I fhall fubjoin a few rules of conduct from Dr. Tissot's ingenious " Advice to the People."

But before I proceed further, I muft again excite the feelings of officers, by declaring, that even in the recovery of health, as well as in the prefervation of it, foldiers have a great deal to expect from their care and affiftance.

The meafures of the furgeon muft be countenanced and enfor-ced; and whatever is wanted and recommended fhould, if poffible, be granted. When men are dif-charged from the hofpital, it cannot be fuppofed that they are entirely reftored to their natural ftrength : That remains to be per-fected by a regulated proportion of ufual exercife and employ-ment; and here they muft truft to the knowledge and humanity of their officers. In circumftances of this kind, the generous and bene-volent will always give every al-lowance; will procure neceffary indulgencies; and in all poffible cafes will find out, and endeavour to obtain, a change of fituation.

H 4 WHERE

WHERE officers have it in their
power to be of ufe to foldiers´ in
the periods of recovery, will be
afcertained in the courfe of the
fucceeding obfervations. I can
here frame in my own mind a part
of that fatisfaction and pleafure
which officers will receive, in
knowing that there are other occafions befides thofe an enemy gives,
of difplaying a wifh to promote
the interefts of their country.

CHANGE

CHANGE *of* AIR.

THE advantages of a change,
particularly from unhealthy
fituations to others of a fuperior
degree of healthinefs, are not
confined, but arife and are evident in all countries. While a
difeafe exifts, good effects are often derived from a change of fituation, even although confidered
in only an equal degree of health
from that which has been forfaken.
Thefe good effects are more certain, if the difeafe has been produced by caufes peculiar to the
fituation—fuch as the vapour of
marfhes : in this cafe, every advantage

vantage may be expected from a change to a situation where thefe caufes do not appear.

As I do not mean to extend my obfervations to the effects of a change of air in the removal of difeafes, I fhall go no further in the preceding explanation, but proceed to confider its effects on patients recovering from difeafe. However, I may previoufly remark, that if a change of air can produce the beft effects, in certain circumftances, on patients labouring under a difeafe; it is reafonable to expect equal and more permanent effects from it, in thofe cafes of which I am to treat.

In

In what manner a change of air, even to an air of no fenfible difference, acts in promoting a reftoration of health, will be unneceffary to inveftigate. Indeed, it would prove a fubject infinitely above the reach of any exertion I could give it : befides, it is only of importance to afcertain by facts, how far the falutary effects of a change of air are to be depended upon, and in what cafes thefe effects will be moft probably produced ; I fhall therefore confine myfelf entirely to this latter enquiry.

THE good effects of a change of air on perfons recovering from difeafe,

difeafe, are amply proved in thofe
patients who remove from St.
Lucia and Tobago to Barbadoes.
I have feen repeated inftances of
perfons with quotidian, even remit-
tent fevers, and many others, who
have been fo much debilitated as
to make their landing attended with
danger, recover beyond concep-
tion, and regain almoft an Euro-
pean degree of health : and, what
may appear fingular, I know cafes
where even a change from Barbadoes to St. Lucia, has been followed by recovery and reftoration
of health.

WHEN every change of air
which the Weft-Indies can afford has been unfuccefsfully
tried, a trip to the continent of
America, or, what is preferable, to
Europe,

Europe, has in feveral inftances effectuated a recovery; whilft a continuance in the Weft-Indies, in all probability, would have proved fatal.

MORE limited changes of air than either of thofe I have fpecified, have often been followed by advantage. Perfons who are attacked with difeafes in the Carenage at St. Lucia, or in any fituation near it, find benefit from removing to fituations about Scouffrier, and to windward of it, and about Gros lflet. The 46th regiment lay for a confiderable time on board of tranfports in Carenage bay, and became very fickly; but on removing to Grofs lflet bay, the ficknefs decreafed, and the fick men gradually got better. The

crews

crews of the Ajax and Vigilant line-of-battle fhips likewife were exceedingly fickly in Carenage bay; but on removing to Grofs Iflet bay the ficknefs abated, and the men very foon recruited.

At Barbadoes, the inhabitants of Bridgetown, on recovery from difeafe, find it their intereft to go into the country; particularly that part of it called Scotland, which is a hilly ground, has its air chiefly from the fea, and is fanned by a conftant wind. In Antigua the inhabitants alfo recover much fooner, by changing their fituation into more elevated and expofed ones.

In

In thofe changes of air, as from
St. Lucia to Barbadoes, a great
deal is attributed to the passage.
The happy effects which generally
accompany the fea-air, or the agi-
tation of the veffel, have long been
demonftrated; and I readily grant
that the passage, though fhort from
one ifland in the Weft-Indies to
another, may have a confiderable
fhare in the recovery which is afterwards perfected. I was very
fenfible myfelf of the benefit
of a voyage, after my leaving St.
Lucia, in the month of July 1779.
Befides my own cafe of a bad ftate
of health, I had with me about
twenty Artillery-men, flowly recovering from difeafe. From St.
Lucia we paffed by St. Vincent,
lay one day off Grenada, then
steered

steered for St. Chriſtopher, which,
until our arrival at that iſland, took
up about fourteen days : during
the paſſage, the weather was clear,
and we generally had a good
breeze of wind. Before we landed
at St. Chriſtopher, I perceived the
happy effects of the voyage, and
we were all in a ſhort period re-
ſtored to very good health.

As the good effects of a change
of air, and of a ſea-voyage, are
placed beyond a doubt, they ſhould
in all poſſible caſes be attempted.
If a diſeaſe is formed by cauſes peculiar to any ſituation, a removal
from it ought to take place immediately, without regarding the
ſtate of the diſeaſe, or the period
of recovery. ·

Dr.

Dr. Lind fays *, that " pre-
" fervation and certain recovery
" depend upon an immediate
" change of air, when feized with
" the prevailing ficknefs of the
" country." He goes on : " I affert
" it as a certain truth; which I
" have had the moft ample means
" of knowing; that perfons labour-
" ing under fevers, fluxes, and
" other difeafes, may with great
" fafety be moved from one place
" to another; nay more, that by a
" removal of them with proper
" care, from an impure to a pure
" air, fuch patients received imme-
" diate benefit.   Remove them
" from the main caufe, and per-

* Advice to Europeans, page 179 and 180.

I   " haps

" haps the only fource of their
" ficknefs ; that is, from the land
" air".

In unhealthy iflands, the Army
would derive many advantages
from being provided with places
fixed in the moft healthy fituations,
when the neceffity of fervice
obliged it to occupy thofe parts,
perhaps the moft unhealthy. If a
provifion of this fort was made,
which with very little trouble
might in almoft every cafe be
done, the fick would be removed,
and have a better chance ; at any
rate, places for convalefcents fhould
always be felected. In fuch an
ifland as St. Lucia, which has a variety of bays, of different degrees
of healthinefs, and feldom defti-
tute

tute of Government tranfports; it
would prove a ftep of the higheft
utility, to have fome of thofe fhips
fitted up and ftationed in the
healthieft bay, which we fuppofe
to be that of Grofs Iflet, for the
reception of the fick of the army;
at leaft, of its convalefcents.

In a former place I mentioned
the probable good confequences
which would accrue from placing
men, immediately after their arrival
from Europe, in the moft healthy
iflands, before they were fent to
thofe deemed very unhealthy; and
I here repeat it, becaufe it gives a
body of men impaired by difeafe an opportunity of being relieved, and removed to a more

I 2   healthy

healthy fituation, by which they will enjoy the effects to be expected from a voyage, and a change of air.

OFFICERS who can obtain leave to go from one ifland to another, fhould delay no time, but immediately adopt a change of air. There are many cafes where officers by delay have fuffered, and there are many cafes where they have narrowly efcaped death. In the prefence of fever, if its nature and the want of conveyance will not allow a change of air, the earlieft opportunity after fhould be embraced. An officer ought not to be refufed leave to remove to any proper place for the recovery

of

of his health. The commanding
officer who objects, from any pre-
tence whatever, is truly charge-
able with every confequence. An
officer during ficknefs cannot do
his duty ; give him then every
chance to recover, and he returns
with cheerfulnefs and gratitude.
If, however, he is detained, and un-
fortunately dies ; the man who was
the caufe of it, has many heart-felt
reflections to encounter. I have
reafon to believe thefe confidera-
tions to be rather impertinent, for
I fpeak of a Britifh commander :
however, the concern I feel in the
diftant thought that it is poffible a
refufal might be given, is the only
apology I offer.

I 3   An

An officer, when he obtains permission to change his situation for the recovery of his health, ought, if it is confined to the same island, to select that place esteemed the most healthy by the inhabitants; and this may be determined also from those circumstances marked in the Observations on the Preservation of Health under the article of Situation. If the leave extends to a removal to another island, as from St. Lucia to Barbadoes, the most healthy situations of the latter should be likewise undoubtedly selected; but it often happens that patients are contented with the mere change to that island, and fix themselves in Bridge-town, the most unhealthy part of it.

PERSONS

PERSONS very much debili-
tated by difeafe muft be cautious of chufing too elevated and
expofed a fituation, where the air
may be piercing and cold. In this
cafe, the fheltered fituation open to
the fouth is the inoft commendable ; and as recovery advances, a
keen air may be lefs guarded
againft. Every degree of air
fhould be gradually received, until the conftitution is enabled to
feel with advantage and fafety the
moft penetrating ftate of it which
the Weft-Indies afford.

IT has been often found, that
all the changes of air obtainable
in any part of the Weft-Indies,
have proved ineffectual in procur-
ing a re-eftablifhment of health.

I 4 · Experi-

Experience, as we have already
obferved, has demonftrated, that
a return to Europe has completed
the recovery, which had baffled
the moft falutary influence of thofe
countries. Officers therefore, after a fruitlefs trial of removing
from one neighbouring place to
another, fhould make a more diftant change, and have leave of
abfence to return to their native
fhore; and it would not be incompatible with the fervice to
extend this leave to the foldier.
Soldiers who continue long in a
convalefcent or recovering ftate,
contract pectoral complaints, fwel-
lings of the legs, &c.: thefe,
however, may be the caufes of an

uncertain

uncertain and tedious recovery, as well as the confequences of it. In any of thefe cafes, there is very little probability of a cure being effected in the Weft-Indies ; the only chance is in a return to Europe, which I think may be accomplifhed with eafe, and without the fmalleft detriment to the fervice ; on the contrary, the moft certain advantages would refult from it.

A REGIMENT may have permiffion to fend once a year thofe men who are deemed by the furgeon irrecoverable in the Weft-Indies to England ; not as invalids or garrifon men, but as men who may recover and return to their

corps

corps, or be drafted into other regiments, as circumftances may
point out. By this method many
men would be annually faved, and
the ftrength of each regiment be
better afcertained.

THERE are regiments which return from five to fix hundred men,
and out of that number more than
one hundred are probably marked
Convalefcent, Confumptive, or Sore
Legs ; and in this cafe the regiment
is defective in that number in
ftrength, befides the incumbrance
given in cafe of fervice : whereas,
by fending thefe men home, the
regiment ftands the chance of re-
ceiving drafts or recruits from England ; by which means it would
be

be always really ftrong, inftead of
being nominally fo; and a number
of brave men would be faved, and
reftored to the fervice of their
country. Sore legs in the Weft-
Indies, in whatever manner they
may be produced, are very troublefome, baffling every effort that
can be made to heal them; and
the hope of their cure can be
built only on the return of the
patient to Europe.

DIET.

D I E T.

NEXT to a change of air, a great deal may be expected, and certainly obtained, in effecting recovery, from a proper attention to what is comprehended under the article of Diet. On recovery, the appetite becomes keen, and not to be very eaſily ſatisfied ; and if this is indulged, which is frequently the caſe, the recovery is retarded, and probably a particular weak ſtate of the ſtomach with other complaints ſupervene. The ſmalleſt degree of exceſs in eating, drinking, or in any other thing, is always accompanied with the

moſt

moſt eminent riſque of producing a relapſe, or laying the foundation of chronic diſeaſes ; it therefore requires a conſtant perſeverance in a regulated diet, and a forbear-ance from every irregularity, to enſure the reſtoration of health.

> While the vital fire
> Burns feebly, heap not the green fuel on ;
> But prudently foment the wandering ſpark
> With what the ſooneſt feels its kindled
>   touch :
> Be frugal even of that ; a little give
> At firſt ; that kindled, add a little more ;
> Till, by deliberate nouriſhing, the flame
> Reviv'd with all its wonted vigour glows.
>
> ARMSTRONG.

In the periods of recovery, thoſe articles of diet of the moſt eaſy digeſtion, and which afford the beſt nouriſhment, ſhould be ſelected.

selected. Even these are to be taken in small quantities; and it ought to be an invariable rule never to take a full meal, but always to desist from eating before the appetite is satiated. In the first dawn of recovery, the diet should consist of liquids or spoon-meats, as broths, preparations of milk, &c. taken in small proportions, and frequently repeated. As recovery advances, those solids the nearest, with respect to digestion, to spoon-meats, may be used; as jellies of the vegetable and animal kinds, young animal meat, and some species of fish. From these we go forward gradually, using substances more solid, and of less easy digestion, until we arrive

at

at our ordinary quantity and kind
of food, when recovery is perfectly effected.

I saw a gentleman at Barbadoes
who came from St. Lucia for the
recovery of his health; he was
very much enfeebled and reduced
by a long-continued attack of fever. For the firſt eight days after his arrival, a fenfible change
took place for the better. During
that time he was confined to the
houfe, except in the evening,
when he took a fhort airing in a
chaife; and he ufed a diet which
had been recommended to him.
Feeling his health and fpirits fo
quickly returning, he became lefs
attentive; and one day, being the
tenth

tenth after his landing, he eat a hearty meal, and drank three or four glaffes of wine : in the evening he took his ufual ride, but complained of a diftenfion of his ftomach. In the night he felt fick, and vomited what he had eaten at dinner; this was attended with headach ; and the next day he had a return of his fever, which continued for fome time ; and with much ado he was refcued from death.

ANOTHER gentleman, in much the fame fituation, but more weakened and reduced, who gave every circumfpection to his conduct, gradually recovered. He prudently declined, and had the

2   refolu-

refolution to abftain from, excefs
and irregularity ; the confequence
of which was, that he was foon
reftored to a ftate of health as vigorous as he had ever enjoyed in
Europe.

. I COULD give a minute detail
of many cafes of recovery, in order to demonftrate the pernicious
tendency of an unguarded indulgence in point of diet, and the
happy effects of a different fyftem ;
but I prefume that what has been
fpecified, and the obfervations
which every one has had opportunities of making, will render it
unneceffary : I fhall therefore proceed to give a few directions re-
fpecting the diet of thofe whofe

recovery is accompanied with fome
particular complaint.

THERE is always a general feeblenefs and weaknefs of the body,
which continue fome time after a
fever has difappeared; and thefe
are in proportion to the nature and
duration of the difeafe. It often
happens that the general weaknefs
remarkably affects fome particular
parts, and none fo commonly as
the ftomach and its dependencies.
As a weak ftate of the ftomach
undoubtedly renders digeftion defective, and nutrition imperfect,
it is impoffible the general ftrength
can be reftored whilft that complaint remains. In cafes of recovery therefore, where it occurs,
the

the principal aim fhould be to remove it. This weak ftate of the
ftomach produces other circumftances, which, united, occafion a
very tedious and troublefome recovery : thefe are, acidity or four-
nefs, and flatulence or wind. Here
the affiftance of medicine muft be
called in : bathing and exercife
will likewife contribute their fhare.
But in all cafes requiring the ufe
of medicine, the furgeon fhould
perfonally direct; and I referve
to a fubfequent place the confideration of bathing and exercife.

In weak ftomachs it is a ftanding rule, to eat little at a time and
often, and to let what is eaten be
of the moft eafy digeftion and

the

the moſt nouriſhing nature. In cafes of this kind, if the ſtomach is too much diſtended, its weak-neſs is increaſed, and every other attendant complaint is aggravated. Milk ; broths ; eggs taken warm from the hen, and eaten either raw or ſoft-boiled ; preparations of well-fermented bread ; vegetable and animal jellies, as ſago, ſalep ; calves heads and feet ſtewed ; tur-tle ſoup, &c. are nouriſhing, and of eaſy digeſtion. As the ſtomach regains its tone, more ſolid ſub-ſtances may be uſed, making a gra-dual progreſs, as I obſerved be-fore, until its natural ſtate is re-ſtored.

SOURNESS

Sourness and wind, in weak stomachs, are produced by an imperfect digestion and assimilation of the food ; therefore, the means of removing them are those which give the usual powers to the stomach. When this impaired digestion is accompanied with these circumstances, all vegetable acids, and bodies readily producing acidity, should be avoided. Animal preparations of every kind are the least liable to give acidity. Dr. Arbuthnot, in his Treatise on Aliments, points out the following vegetables as anti-acid, viz. cabbage, turnips, carrots, onions, leeks, radishes, and mustard. Rum mixed with water makes the best common liquor. Wine in

every

every case of weakness is one of
the most effectual articles we are
possessed of, if used moderately
and without intemperance * : however, in weak stomachs the greatest delicacy is required in using it,
for it sometimes produces sourness. Red-wines are the most un-
exceptionable, and Port the least
so of any. Tea, or indeed any
warm infusion, is hurtful to weak
stomachs; every thing liquid
should be taken cold : and as a
substitute for tea, I would recommend a cold infusion of chamomile
flowers, sweetened, and coloured
with milk, which after short use
becomes agreeable.

* Vide the first part on Diet.

Costive-

Costiveness is frequently troublefome in the periods of re- covery ; and if there is no acidity in the ftomach, a vegetable diet is the moft commendable. Figs, raifins, pruens, grapes, oranges, tamarinds, honey, and cream-tartar, and water, may be occafionally ufed. In cafes of acidity, a tea- fpoonful or two of magnefia will produce an effect. If thefe how- ever fail, recourfe muft be had to more active laxatives.

Looseness fometimes retards recovery. When it is flight, and proceeds from too great acidity in the ftomach, abftinence from vege- tables, and fmall draughts of lime-water, or a little magnefia,

K 4   taken

taken occafionally, will in general remove it. Loofenefs is a complaint that often continues, and proves dangerous; therefore, in all cafes where it remains more than twenty-four hours, application fhould be made for medical affift-ance. In cafes of coftivenefs, white-wines are the beft; and in loofenefs, the red-wines, of which Port is the moft aftringent, are recom-mended.

A SWELLING of the legs and ancles generally attends recovery, in a greater or lefs degree, according to the ftate of weaknefs and relaxation; it goes off as recovery advances, and difappears when that is perfect. The feet and legs
fhould

should be frequently rubbed with a flesh-brush; and if the swelling is considerable, and gives pain, a little oil with camphor or spirit of hartshorn may be used previous to the friction. If this swelling is attended with a more extensive dropsical disposition, the assistance of medicine becomes necessary.

THERE are other complaints besides those I have mentioned, that follow the diseases of the West-Indies, and make recovery from them tedious and uncertain : but as they always require personal examination and advice, I omit giving any description of them.

THE preceding observations on Diet are principally directed to officers ; but from them several things

things may be selected, as applicable to the soldier. Convalescents, or recovering soldiers, continue under the direction of the surgeon; but when he thinks it proper to discharge them from the hospital, they must not be supposed to be perfectly restored to their former strength. The entire restoration of their health remains to be effected by a gradual return of their usual mode of life, which it is to be presumed undergoes the regulation and inspection of their officers.

CONSIDERABLE advantage, I think, would result from classing the men discharged from hospitals into separate messes, and bestowing more

more particular attention on their diet. If frefh provifions cannot be obtained, the falt beef fhould be foaked in warm water; which, frequently repeated, would extract its faltnefs, and make it more eafily affimilated to the purpofes of nutrition : and inftead of boiling it in the common manner, it fhould be cut into flices, and ftewed with mucilaginous vegetables.

SOLDIERS for fome time after their removal from the hofpital, fhould have an allowance of wine inftead of fpirit, and in every poffible cafe frefh meat. Government, befides permitting this alteration in diet, would find advantage in giving a pint of porter per day to each man

in

in the fituation I have fpecified;
and I venture to affert, that the
increafe in the pecuniary charge of
the ration, would be very trifling,
by honeft management. In recovery, porter is a pleafant and ufeful
liquor, and may be taken in moderation at any time when wifhed
for; unlefs it difagrees with the
ftomach, which it fometimes does
in cafes of acidity.

DRESS

DRESS *and* EMPLOYMENT.

DURING the day, the drefs
of perfons recovering from
difeafe may be loofe and light ;
but where the weather is damp,
rainy, or variable, woollen cloaths
are the beft. Perfons much enfeebled receive very great advantage
from a flannel fhirt worn next to the
fkin : at firft this may feel difagreeable, but a few days wear-
ing will reconcile it. By wear-
ing a flannel fhirt, a thinner
exterior cloathing can be ufed,
which otherwife would be improper : befides this, the flannel has
the effect of retaining natural
heat,

heat, the fmalleft degree of which,
in cafes of recovery, is valuable.
It has another advantage, of preventing the unpleafant effects of perfpiration, which linen always imparts.

Soldiers, after their dif-
miffion from the hofpital, fhould
be for fome time permitted to wear
only their regimental waiftcoats and
breeches, with woollen and cotton
ftockings; and this ought to be
more particularly attended to in
night duty, or in any other expofed fituation. There may be a
neceffity for putting foldiers too
early on duty after their recovery:
in fuch cafes, confiderable benefit
would be derived from, and bad
confequences prevented by, the ufe

of

of the flannel fhirt, which ought to be provided at the expence of Government; but if that cannot be granted, commanding officers would find advantage from making it an individual expence, rather than it fhould not be obtained.

I AM aware of the objections which have been made to the flannel fhirt, from the fuppofed danger of changing it, or leaving it off altogether; but this will be found, upon trial, only imaginary. Weaknefs characterifes recovery, which advancing the conftitution regains its natural ftrength. When the conftitution is perfectly reftored, no prejudicial confequences will follow the laying afide of the flannel fhirt: until that, however, is effected,

it

it may be proper to retain it. I
have seen good effects in the cases
I have mentioned from the use of
it ; but I have never been able to
observe any bad effects from its
disuse, after recovery was complete-
ly effected.

Soldiers should have more
attention paid to them, to cleanliness, and neatness of dress,
after they are discharged from
the hospital, than at any other
time ; because diseases are apt to
leave a languor and lowness of
spirits, which beget, if not prevented, dirtiness and indolence.

The observations under the
article of Employment, on the
means of preserving health,
may be partly introduced here ;

3

parti-

particularly thofe which relate to going to bed early, and rifing early, and to maintaining an active and cheerful ftate of mind. In every cafe of recovery, good and well-timed fleep affords fenfible advantage : indeed, we can from the foundnefs of fleep often date the certainty of recovery, and even the period when that will happen. By it all the natural powers are ftrengthened; being relieved from their ordinary action, they return to it with an increafe of vigour. From rifing early, the pleafing fenfations of a cool air, and the proper time of exercife, are obtained ; befides, the bad effects of morning fweats are prevented. There is nothing that retards recovery more

L   than

than profufe perfpiration ; it not only proves the degree of relaxation and weaknefs, but always increafes it. This injurious perfpiration moft frequently occurs early in the morning ; in which cafe, it fhould be an invariable rule to forfake the bed inftantly, and have the fkin well rubbed with a dry cloth, or a flefh-brufh. It is beft to rife, be it ever fo early ; even though, by doing fo, it may be neceffary to lay down for a fhort time, fome hours after.

Exercise of every kind is an effential circumftance in effecting recovery, and it fhould be proportioned to the feelings and ftate of ftrength. The mornings and evenings are the only proper periods of

of ufing it. The fun fhould make
his appearance, before riding or
walking is attempted, and either
may be continued one or two hours;
and the evening exercife fhould
commence about two hours before
his fetting, and terminate almoft
immediately after it. By this
mode you avoid the damp atmofphere of the morning and evening, which ought always, but more
particularly in cafes of recovery,
to be fhunned and guarded againft.

SOLDIERS who cannot ufe the moft
eligible exercife, have a good fubftitute in the morning and evening
manual manœuvres. Soldiers when
they are difcharged from the hofpital fhould be marched out, and

exer-

exercifed every dry morning and evening, until they have entirely regained their former ftate of health. They fhould not be fent too early on any expofed duty ; the commanding officer can always employ them in fome way in which they will not feel the viciffitudes of weather.

In every period of recovery, and particularly in the early ftages of it, expofure to the fun, except at the hours I have mentioned, fhould be carefully avoided ; but if ne-ceffity will not allow this indul-gence, the directions for preventing its effects ought to be more punc-tually attended to. I have often feen relapfes, and even death, follow an

2 impru-

imprudent continuance and exercife in the fun, in cafes of recovery.

BATHING in the fea, in rivers,
in domeftic cold baths, or by the
fimple procefs of having two or
three pailfuls of cold water thrown
upon the body, is of the higheft
importance in facilitating recovery.
In the firft ftages of recovery, the
furgeon fhould be confulted ; for
there are certain circumftances
which may render bathing improper, or require a great deal of circumfpection in its ufe. If medical affiftance cannot be obtained, it
fhould be a rule never to bathe in
cold water, whilft any particular
complaint exifts : for example, a
headach, cough, pain in the

L 3   breaft

breaſt or bowels, or any ſimilar affection. After bathing, if any diſagreeable feelings ariſe and remain, it ought to be diſcontinued : but if it is followed by active and cheerful ſenſations, evidence is given that the beſt effects may be expected from it. Early in the morning is the fitteſt period to bathe, and I would prefer the ſhore-bath, or the throwing two or three pailfuls of water over the body. The ſhock by theſe methods is greater than by the others, and it may be increaſed or diminiſhed at pleaſure. When a chillineſs continues ſome time after bathing, it ſhews that the ſhock has been too great ; in which caſe it will be proper to lie down in bed, and

3   drink

drink of any warm liquid. The
pleafing glow which fucceeds the
healthy bathing, may be generally obtained by the preceding
mode of returning into bed and
drinking fomething warm, or by
rubbing immediately after bathing
with a flefh-brufh. In all cafes
where bathing is ufed, I recommend rubbing with a brufh, or a
piece of flannel, until the external
part of the body becomes agreea-
bly warm.

SOLDIERS, during recovery, never fhould be permitted to bathe
without the direction and infpection of the furgeon ; and when they
return to duty, they ought to do it
under his eye, or the prudent and
watchful conduct of their officers.

THE

THE difeafes of the Weft-Indies are always accompanied with a particular depreffion of mind, which is very apt to continue, and will certainly retard recovery. An anxious and low ftate of the mind is always produced by, or connected with, a weaknefs, or an impaired frame of the body ; and this fituation of the mind, inftead of being merely an effect or attendant of the valetudinarian habit, becomes an active caufe and ftrong fupport of it : it therefore fhould be a conftant aim to be lively and cheerful

OFFICERS are apt to indulge the wifh of getting to Europe ; and if they folicit for permiffion and are refufed, they become dejected and fretful

fretful, circumſtances which op-
poſe the reſtoration of health. On
ſuch occaſions, that manly forti-
tude and perſeverance which
are the characteriſtic qualities of a
ſoldier, ſhould be peculiarly exerted. The converſation of a
friend, the ſprightly humour of a
ſocial company, and the virtuous
indulgence of that inclination
which leads the Britiſh officer into
the female circle, ought to be alter-
nately adopted.

———————

I AM now to ſubjoin, with oc-
caſional Notes, a few Obſervations
and Rules ſelected from Dr.
TISSOT's ingenious "Advice to the
People;"

People ;" publifhed and tranflated by Dr. KIRKPATRICK, in the year 1771.

THE term of recovery from a dif-eafe requires confiderable vigilance and attention, as it is always a ftate of feeblenefs, and thence of depref-fion and faintnefs. The fame kind of prejudice which deftroys the fick, by compelling them to eat during the violence of the dif-eafe, is extended alfo into the ftage of convalefcents, or recovery ; and either renders it troublefome and tedious, or produces fatal relapfes, and often chronical diftempers. Whenever the fever is compleatly terminated, fome different foods may be entered upon : fo that the pa-tient

tient may venture upon a little white-
meat, provided it be tender ; fome
fifh ; a little flefh foup ; a few eggs
at times, with wine properly diluted.

It muft be obferved at the fame
time, that thefe very proper aliments, which reftore the ftrength
when taken moderately, delay the
perfect cure if they exceed in
quantity, tho' but a little ; becaufe
the action of the ftomach, being
extremely weakened by the difeafe
and the remedies, is capable only,
as yet, of a fmall degree of digeftion ; and if the quantity of its
contents exceeds its powers, they
do not digeft : frequent returns
of the fever fupervene.

EVERY

EVERY bad confequence is prevented by the recovering fick con-
tenting themfelves, for fome time,
with a very moderate fhare of proper food. We are not nourifhed in
proportion to the quantity we fwal-
low, but to that we digeft.

A PERSON on the mending hand
who eats moderately, digefts it, and
grows ftrong from it. He who
fwallows abundantly, does not
digeft it; and, inftead of being
nourifhed and ftrengthened, he
withers infenfibly away.

RULES.

R U L E S.

1. LET thofe who are recovering take very little nourifhment at a time, and take it often.

2. LET them take but one fort of food at each meal, and not change their food too often.

NOTE. This rule I think too abfolute; a perfon in recovery may with care indulge his tafte of variety in fubftances of eafy digeftion, and of a nourifhing quality.

3. LET them chew whatever victuals they eat very carefully.

4. LET

4. LET them diminifh their quantity of drink. The beft for them in general is water, or toaft and water, with a fourth or third part of white wine. Too great a quantity of liquids at this time prevents the ftomach from reco-vering its tone and ftrength; it impairs digeftion, &c.

NOTE. An exception may be made to one part of this rule re-fpecting wine, as in cafes of loofe-nefs, and where an acidity or four-nefs prevails, in which I would recommend red, in place of white wine ; and in every cafe I think a glafsful of pure wine may be ventured upon, and repeated ac-cording to the degree of recovery.

5. LET

5. LET them go abroad as often as they are able, whether on foot, in a carriage, or on horseback. This last exercise is the healthiest of all. If exercise is taken soon after a meal, it impairs digestion.

NOTE. The first part of this rule, relating to the mode of exercise, must be conducted in the West-Indies with the greatest circumspection and care; and the periods for exercise which I have formerly pointed out should be chosen.

6. As people in the state of recovery are seldom quite as well towards night, in the evening they should take very little food. Their sleep will be the less disturbed for this,

this, and repair them the more and
fooner.

Note. This is a very important
rule, and ought to have every compliance given to it. Light fup-
pers in a ftate of health are only
commendable in the Weft-Indies.

7. They fhould not remain in
bed above feven or eight hours.

8. The fwelling of the legs
and ancles, which happens to moft
perfons at this time, is not dangerous ; and generally difappears of
itfelf, if they live foberly and regularly, and take moderate exercife.

It

9. It is not neceffary in this ftate that they fhould go conftant-ly every day to ftool ; though they fhould not be without one above two or three days.

10. Should they, after fome time, ftill continue very weak ; if their ftomachs are difordered ; if they have, from time to time, a little irregular fever ; they fhould take fome dofes of bark daily, which fortifies the digeftions, re-covers the ftrength, and drives away the fever.

Note. This is a rule that does not come within the limits of my Obfervations ; it relates to the ufe of medicine, the confideration of

M   which

which I have all along avoided; becaufe I advife the furgeon to be
confulted in every cafe where that
becomes neceffary.

11, and laft. THEY muft by no
means return to their labour or
ufual employment too foon.

THE END.

www.ingramcontent.com/pod-product-compliance
Lightning Source LLC
Chambersburg PA
CBHW020541270326
41927CB00006B/680